4ᵗʰ Grade Science Volume 2

© 2013 Todd Deluca
OnBoard Academics, Inc
Newburyport, MA 01950

800-596-3175
www.onboardacademics.com

Table of Contents

Earthquakes

What is an earthquake?

We generally think of an earthquake as a violent and unexpected shaking of the earth's surface which causes considerable destruction and loss of life. However, this is true of only a few of the millions of earthquakes that occur each year, most of which are too small to be noticed.

What causes earthquakes?

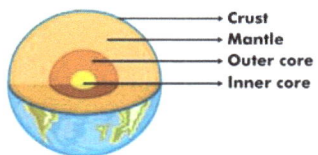

Earth is made up of the inner core, the outer core, the mantle and the crust. The mantle and the crust together form what is known as the lithosphere. The lithosphere is made up of many plates called tectonic plates which are continuously and slowing moving past each other.

Sometimes these plates collide or stretch away from each other resulting in a large release of energy which spreads in waves called seismic waves. These seismic waves are what sometimes cause the very violent shaking of the earth that we know of as an earthquake.

A **fault** is a name given to crack in the surface (crust) of the Earth. Faults occur at the boundaries where tectonic plates meet. In a **horizontal fault**, two blocks of rock on either side of the fault move in opposite directions. In a **vertical fault**, the rock on one side of the fault moves down creating a cliff.

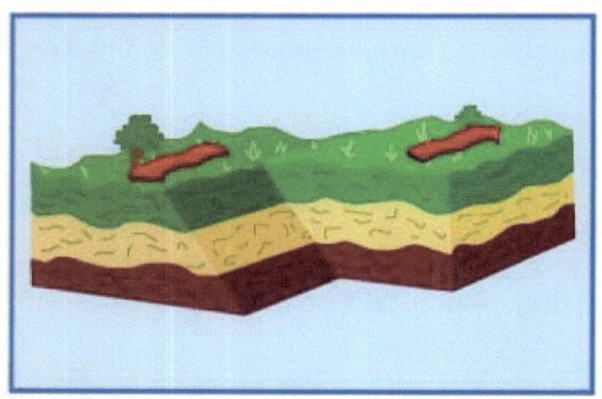

Horizontal Fault

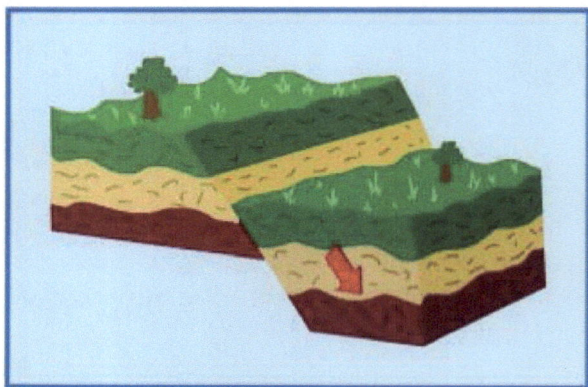

Vertical Fault

The **focus** of an earthquake is the point inside the Earth's crust where the shifting of the rock takes place. This causes a release of energy which, in larger earthquakes, causes the Earth's surface to shake. The point directly above the focus at the surface of the Earth is called the **epicenter** and is where most damage tends to occur.

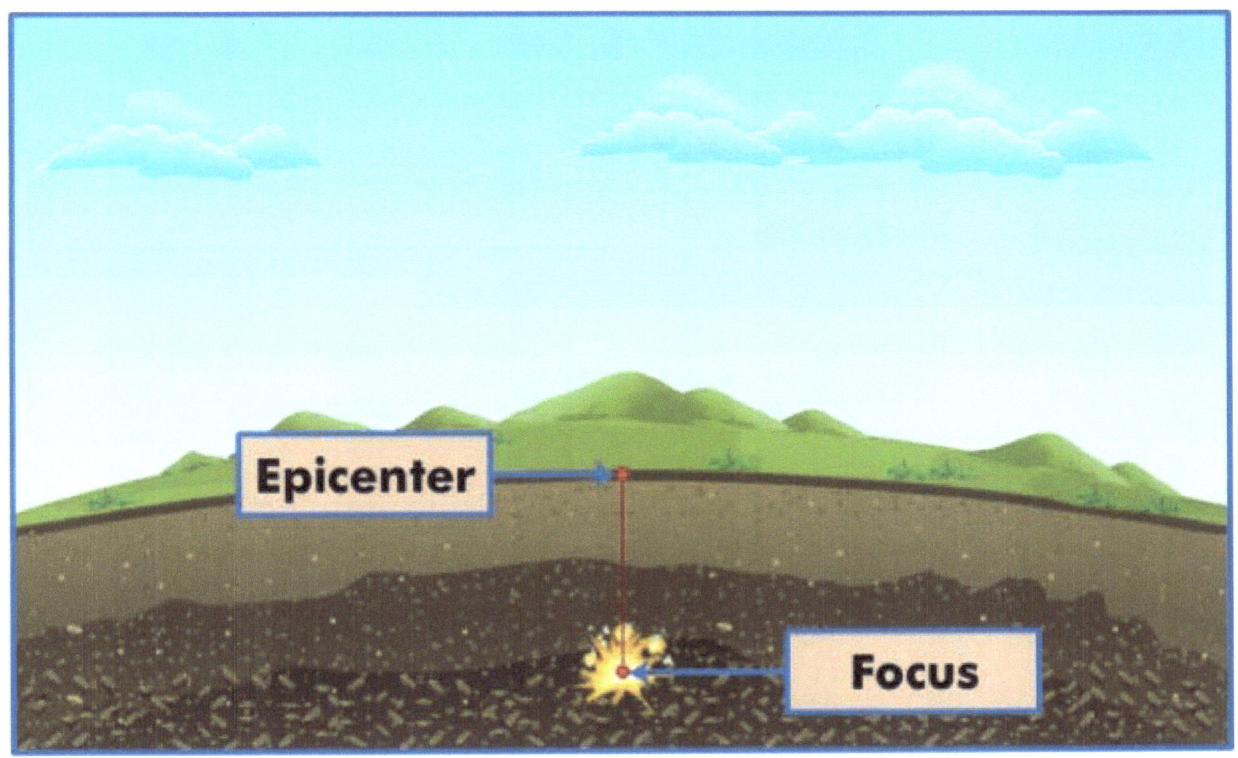

Earthquakes can be measured and recorded using a device called a **seismograph**. In a seismograph, a 'pen' is kept in a stationary position above a rotating cylinder. When the Earth shakes, it shakes the cylinder causing the pen to draw a zigzag line on a roll of paper attached to the cylinder.

The magnitude and intensity of earthquakes is indicated on the Richter scale, a 0-10 scale in which each number indicates a magnitude 10 times that of its preceding number. For example, an earthquake that measures 6 on the Richter scale has a magnitude 10 times greater than an earthquake which measures 5.

Most scientists today use a slightly different scale (the **moment magnitude** scale), and use digital instruments to measure and record earthquakes.

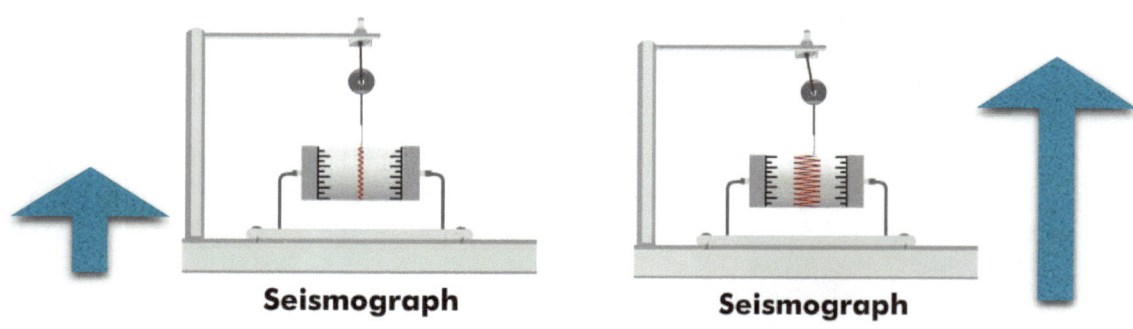

Seismograph   **Seismograph**

Order the Richter magnitude scale definition from 1-10, 1 being the smallest earthquake.

☐	**Very noticable and damage to buildings.**
☐	**Barely detectable and generally not felt.**
☐	**Destructive close to epicenter.**
☐	**Devastating over thousands of miles.**
☐	**Noticable shaking and minor damage.**
☐	**Destructive over a wide area.**
☐	**Destructive over several hundred miles.**
☐	**Never recorded before in human history.**
☐	**Detectable only by instruments.**
☐	**Felt, but minimal if any damage.**

Earthquakes Around the World

Earth's surface is a little bit like a giant jigsaw puzzle. We call the pieces of the puzzle tectonic plates, and earthquakes occur where tectonic plates meet. This is why certain parts of the world are more prone to earthquakes.

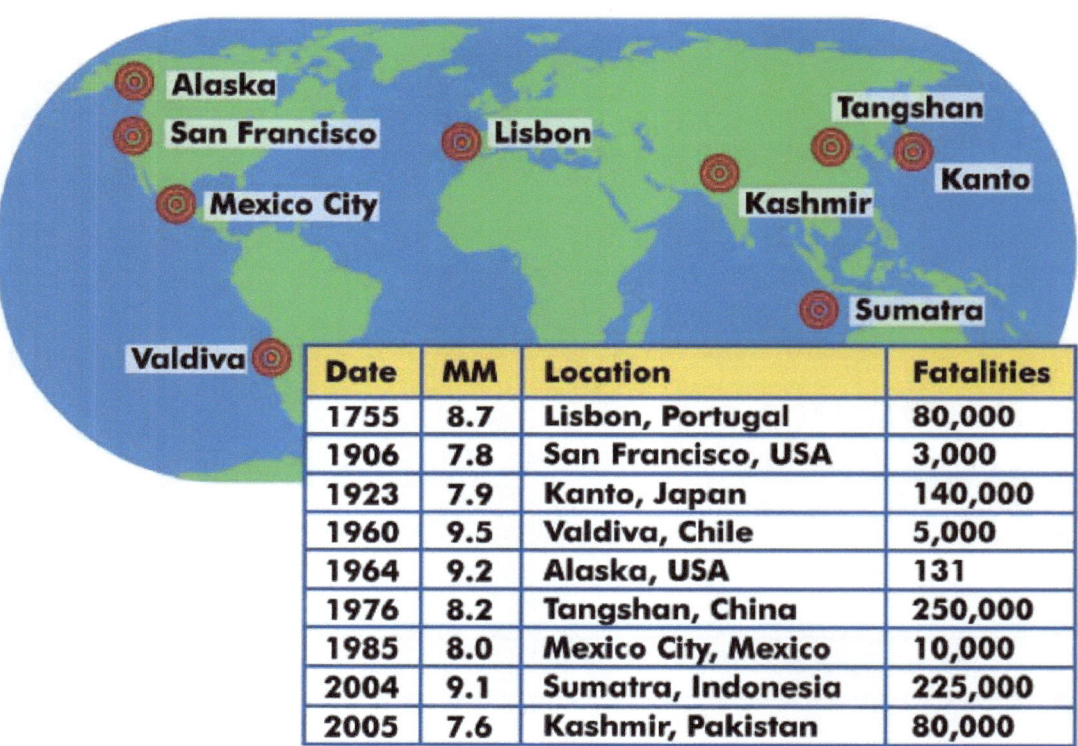

Date	MM	Location	Fatalities
1755	8.7	Lisbon, Portugal	80,000
1906	7.8	San Francisco, USA	3,000
1923	7.9	Kanto, Japan	140,000
1960	9.5	Valdiva, Chile	5,000
1964	9.2	Alaska, USA	131
1976	8.2	Tangshan, China	250,000
1985	8.0	Mexico City, Mexico	10,000
2004	9.1	Sumatra, Indonesia	225,000
2005	7.6	Kashmir, Pakistan	80,000

1. Where did the largest earthquake occur? _____
2. Where was the oldest recorded earthquake on this chart? _____
3. Do the number of fatalities correspond to the size of the earthquake? _____
4. What seems to be the biggest factor regarding earthquakes and fatalities?
   _____
5. Why is did the largest earthquake on this chart cause the least amount of fatalities?
   _____

How are tectonic plates, earthquakes and volcanoes related?

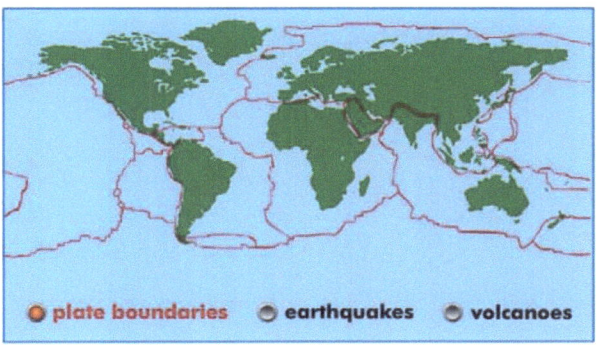

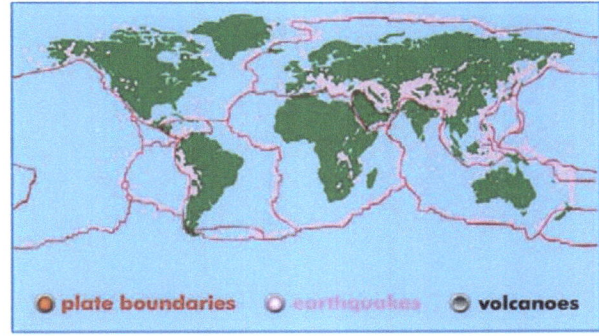

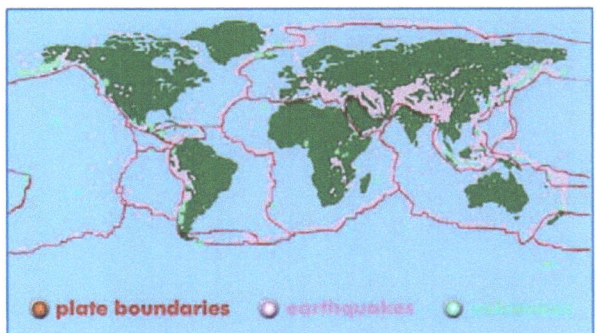

Earthquakes and volcanoes often occur in the same locations: at the boundary where tectonic plates meet. Volcanoes occur when hot magma escapes through a crack in the Earth's crust that has been caused by shifting plates. When the magma erupts through the Earth's surface it's called lava, and the lava cools to form solid rock.

Earthquakes Under the Ocean

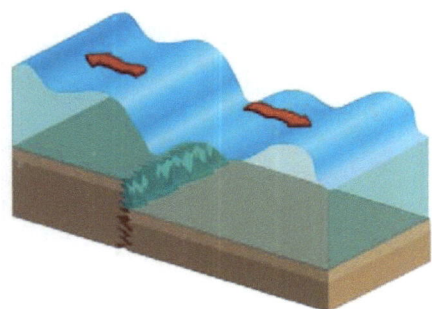

When an earthquake occurs under the ocean it pushes the seabed upwards. This displaces water up to hundreds of kilometers. Starting at the epicenter, large waves move through the ocean at increasing speeds and heights.

The height and intensity increase as they near the coast. This phenomenon is called a tsunami and means tidal wave and can cause considerable destruction and loss of life.

_____

According to the FEMA Guidelines what should you do in the event of an earthquake? Place a √ for yes and an X for no. Answers after the quiz.

INSIDE	Get outside straight away.	
	Take cover under a sturdy table.	
OUTSIDE	Stay outside.	
	Find shelter in the nearest building.	
DRIVING	Park under the nearest bridge.	
	Get out of your vehicle.	
TRAPPED	Move around to attract attention.	
	Tap on a pipe or wall.	

Earthquakes Quiz

1. How long do earthquakes generally last?
   - a. a few seconds
   - b. a few seconds to a few minutes
   - c. an hour
   - d. a few hours

2. The collision of two tectonic plates during an earthquake can create mountains. True or false?

3. The layers of irregular slab that form the lithosphere are called: _____.
   - a. Tectonic plates
   - b. Core
   - c. Mantle
   - d. Crust

4. An earthquake by itself does not cause any harm. Damage is caused only by buildings and trees falling. True or false?

5. The location directly above the focus on the surface of the Earth is called the epicenter. True or false?

6. The Richter scale is used to represent the position at which an earthquake occurred. True or false?

Answers to the FEMA guideline questions.

INSIDE	Get outside straight away.	❌
	Take cover under a sturdy table.	✅
OUTSIDE	Stay outside.	✅
	Find shelter in the nearest building.	❌
DRIVING	Park under the nearest bridge.	❌
	Get out of your vehicle.	❌
TRAPPED	Move around to attract attention.	❌
	Tap on a pipe or wall.	✅

FEMA is the Federal Emergency Management Agency which in 2003 became part of the U.S Department of Homeland Security. FEMA's mission is to support the nation in the event of a disaster such as an act of terrorism or an act of nature such as a hurricane or an earthquake.

Volcanoes

How much of the Earth's surface came from volcanoes?

More than 80% of all Earth's surface (above and below sea level) came from volcanoes.

What is a volcano?

Volcanoes occur when hot gasses and magma from the mantle escape through cracks in the Earth's crust. When magma reaches the Earth's surface it's called lava.

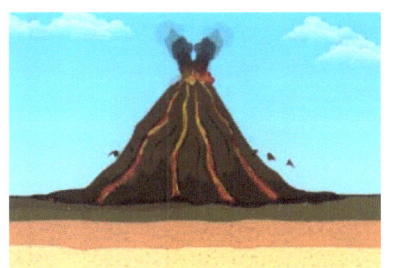

Sometimes Volcanoes explode violently and expel lava, bombs (large chunks of lave, ash and gasses into the atmosphere.

In some volcanoes, lava oozes gently from a vent in the volcano in a liquid flow.

Whatever way lava reaches the surface, lava cools and forms solid rock which helps to form new landforms including mountains and islands. Mt. Hood in Oregon and the islands of Hawaii, Japan and the Philippines were all formed from volcanic eruptions.

Most volcanoes are found in the Earth's two main earthquake belts; the Circum Pacific belt and the Alpine Himalayan Belt. This is where the Earth's tectonic plates come together or move away from each other.   The Circum Pacific Belt is known as the Rim of Fire and is home to about 75% of the Earth's active and dormant volcanoes

Volcanoes occur when hot magma escapes through a crack in the Earth's crust. When magma reaches the Earth's surface it's called lava. Most volcanoes occur where Earth's tectonic plates come together or move away from each other.

Three Main Types of Volcanoes
Read about the different types of volcanoes and then label the illustrations below

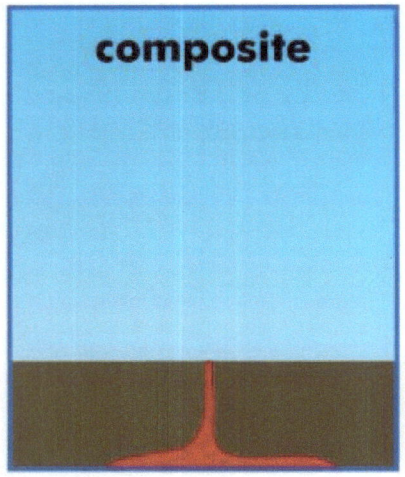

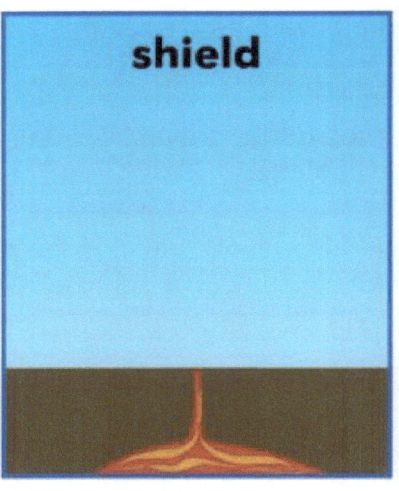

Composite volcanoes also called stratovolcanoes are tall symmetrical volcanoes made of alternating layers of cinder and ash.

They are formed by explosive eruptions followed by quiet periods.

Some of the worlds most famous mountains are composite volcanoes including Mt. Fuji, Mt. St. Helens and Mt. Kilimanjaro.

Shield volcanoes are formed by gently oozing flows of lava instead of large explosions.

Because of that they tend to be very large and quite flat resembling shields. The islands of Hawaii are shield volcanoes.

Cinder Cone volcanoes are the most common form of volcanoes. Their shape is the one we most associate with volcanoes; steeply sloping sides and a bowl shape crate at the top.

Cinder cone volcanoes form fairly rapidly as a result of explosive eruptions. Ash, cinders and volcanic rock are hurled into the air and fall back to earth near the vent.

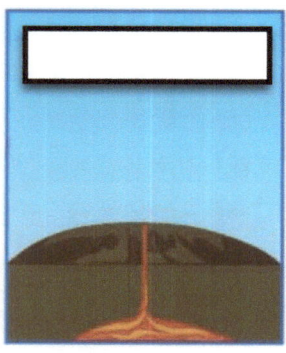

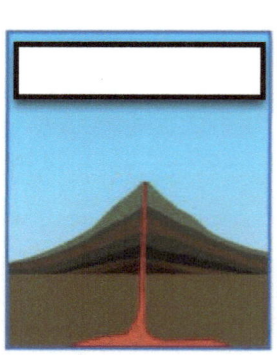

Let's try again to match the volcano type with the brief description.

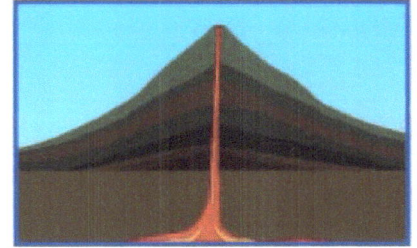

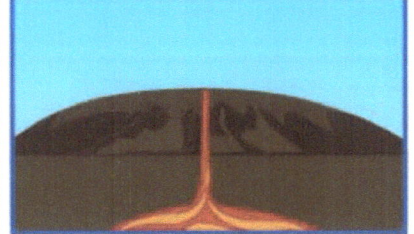

cinder cone   **composite**   **shield**

I'm formed by fluid flows of lava.

I'm formed when ash, cinders and volcanic ash explode and fall back down near my vent.

Sometimes I'm explosive and sometimes I'm quiet. That's why I'm made of layers of lava and ash.

Label the parts of the volcano.

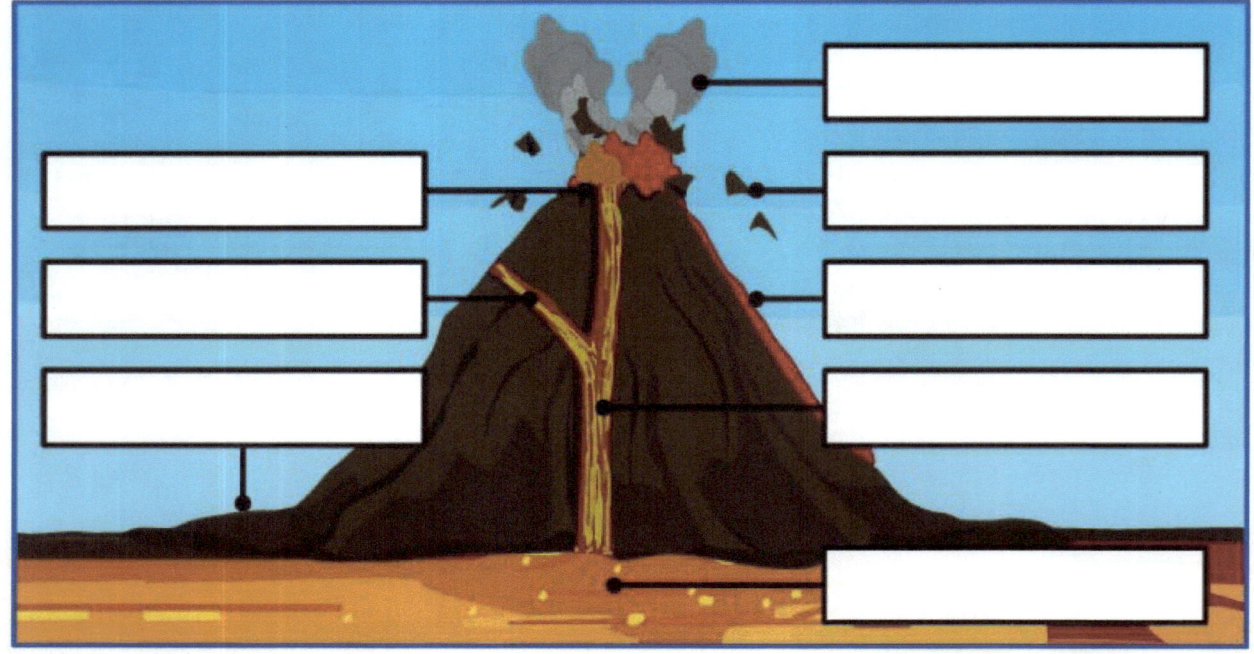

side vent   **crater**   **magma chamber**   **vent**

ash cloud   **lava**   **Earth's crust**   **bomb**

Volcano Quiz

1. More than 90% of the Earth's surface came from volcanoes. True or false?

2. The _____ is the outermost layer of the Earth and the layer which forms the land and the ocean floors.
    a. crust
    b. mantle
    c. core
    d. magma

3. On cooling, _____ forms solid rock which helps to crate new landforms.
    a. magma
    b. lava
    c. the mantel
    d. sand

4. The Circum Pacific belt is also called the _____.
    a. Flame of Fury
    b. Mount Hood
    c. Alpine Himalayan Belt
    d. Ring of Fire

5. Shield volcanoes are also called stratovolcanoes. True or false?

Minerals

What is a rock made of?

Stone   Iron   Water   Steel   Minerals

> **Minerals are the building blocks of rocks which are formed from two or more minerals.**

What is a mineral.

Minerals are inorganic (non-living) substances which are made from one or more chemical elements which form in a *crystalline* structure. This means that the tiny particles that make up the mineral (called atoms) are arranged in a very regular pattern such as a square or a double triangle. There are more than 3,000 different types of minerals on Earth, but only a few dozen are commonly found.

How many types of minerals are found on earth? _____

Do minerals contain living matter? _____

How many minerals are commonly found? _____

How many elements make up a mineral? _____

Where do minerals come from?

Most of our minerals are formed when molten rock, deep within the earth, moves toward the surface and erupts from volcanoes and then cools.

Elements inside the cooling magma come together in organized and repeating patterns to form minerals. A crystal is the name given to a solid block of minerals in this repeating pattern formation. If there is ample space minerals may form very large crystals. Talc is one of the softest minerals and diamonds are one of hardest.

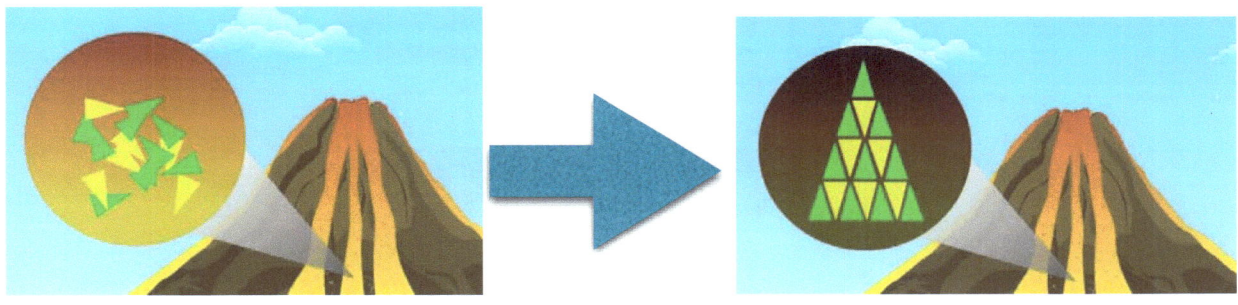

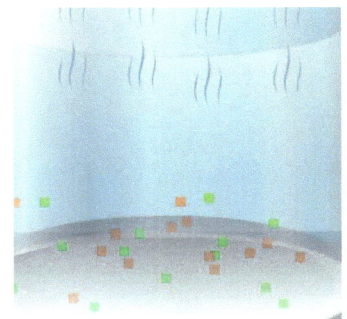

Minerals are also made when water that contains certain chemicals evaporates. For example, if you let salt water evaporate your are left with sodium chloride mineral more commonly known as salt.

Minerals are made when magma (liquid rock) cools into solid rock and also when water evaporates and leaves behind tiny pieces of minerals that were dissolved in the water.

Five characteristics of minerals

Hardness measures what things the mineral can scratch and what can scratch the mineral.

Color describes the mineral's color after you have removed dirt and other materials.

Cleavage describes what the mineral looks like when you break it.

Streak describes what kind of mark the mineral leaves when you try to write with it.

Luster describes if the mineral is shiny like metal or dull.

On a scale of 1-10 how hard is each mineral? Use the hints below.

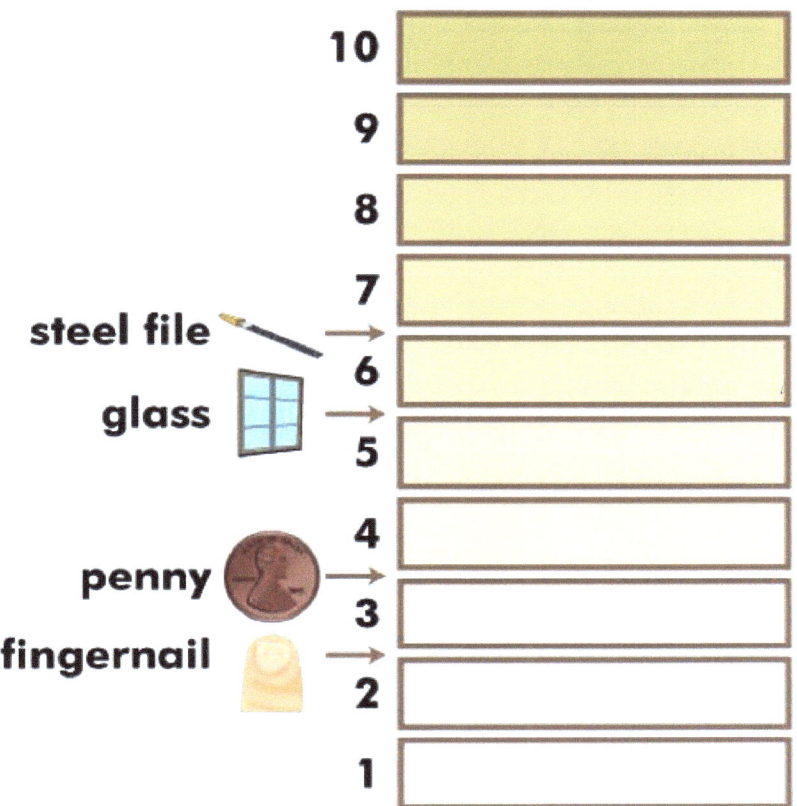

Orthoclase can scratch glass, but apatite cannot.
Quartz, Topaz, Corundum and Diamond are all harder than steel.
You can scratch calcite with a penny.
You can scratch gypsum and talc with your fingernail.

www.onboardacademics.com

Are you a mineral master?
Fill in the blanks below.

**All rocks are made of at least two different types of
_____. Minerals are made when _____ in
the Earth cools off, or when _____ evaporates.
There are five common properties of minerals: hardness, color, cleavage, streak, and _____.
_____ is one of the softest minerals; _____
is the hardest. The same mineral often appears in different colors, so to identify a mineral, you can use
a _____ test which means to rub the mineral on a
tile to see what color it makes. Cleavage is used to de-
scribe what a mineral looks when it is _____.
Luster describes if a mineral is shiny or _____.**

streak   water   Talc   dull

magma   diamond   minerals   luster   broken

Minerals Quiz

1. _____ are the building block in rocks.

2. Minerals are inorganic substances made from a single element. True or false?

3. Minerals are crystalline in nature. True or false?

4. Minerals are made when _____ cools into solid rock.

5. Which of the following is one of the softest minerals?
   a. Quartz
   b. Topaz
   c. Talc

Contour Maps

Can you identify each type of map?

> **Political maps** show national and regional boundaries. **Physical maps** show the physical features of a region, e.g. mountains and rivers. **Topographic maps** (also called **contour maps**) show the surface features of a region using contour lines. Contour lines show elevation: the height of a feature or structure above sea level. Topographic maps also show man-made structures, such as roads and buildings.

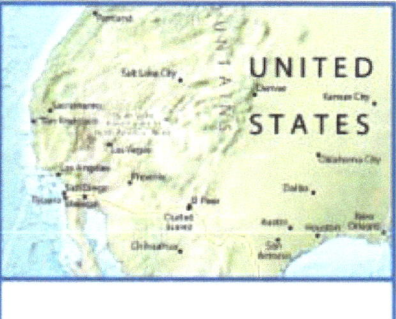

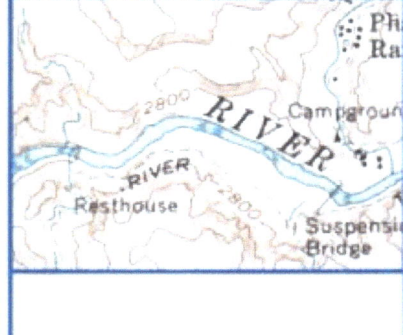

physical   topographic   political

Which is the correct topographical map for this feature?

Contour lines are used to connect points of equal elevation. For this reason, contour lines can never cross, because a feature cannot exist at two different elevations.

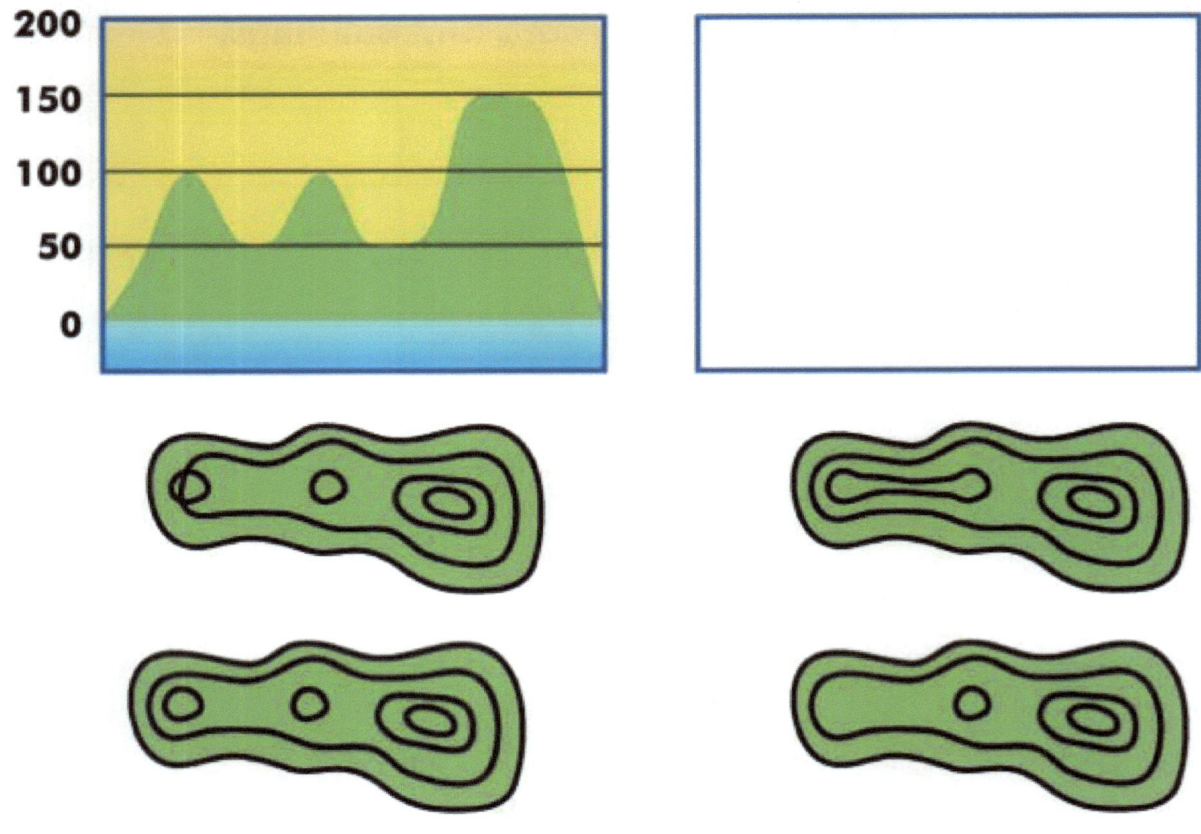

Identify the Great Plains and the Rocky Mountains.

Contour lines can give you an insight into the slope of a land surface. Lines that are close together indicate a steep slope, while lines that are farther apart indicate a more gradual slope.

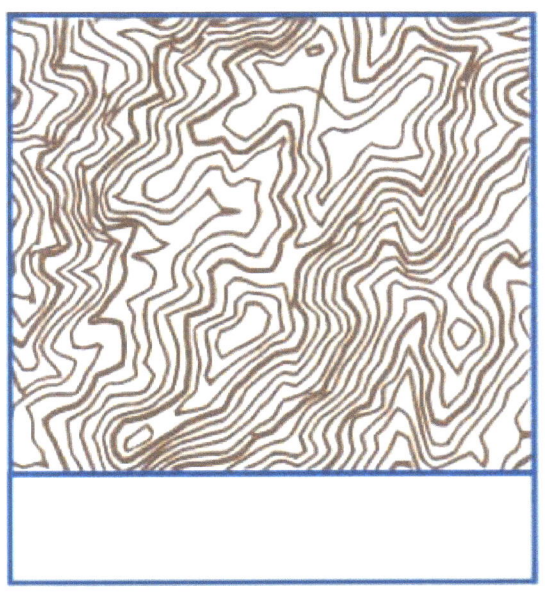

Rocky Mountains   **Great Plains**

Identify the depression and the river.

Label each map either depression or river.
Copy the color dots from each map into the boxes to represent the highest and lowest points on each illustration. If you don't have color, write the color's name.

> **Topographic maps show depressions as well as hills and mountains. Depressions are represented by contour lines with little marks that point inward. Contour lines that cross a river form a "V" which points towards higher elevation.**

□ □ **higher elevation**   □ □ **lower elevation**

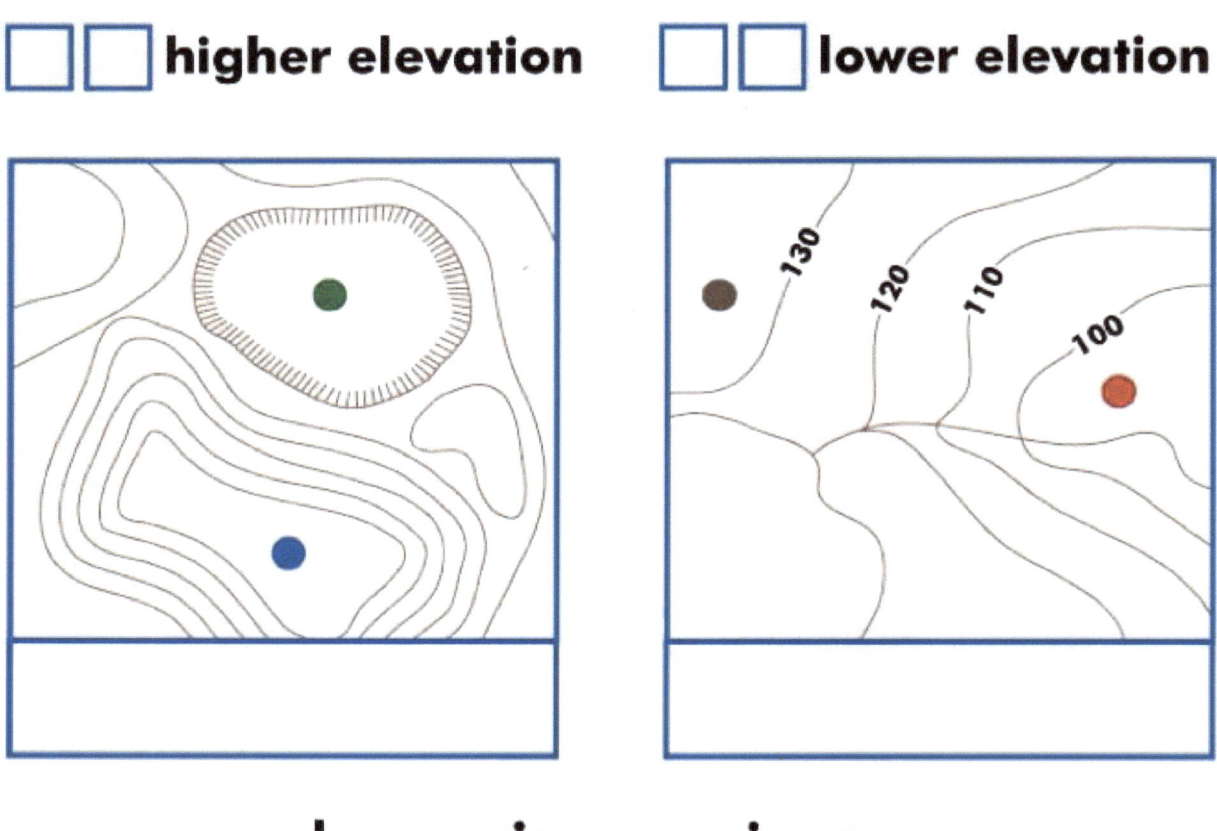

depression   **river**

Match each contour map with the correct profile.

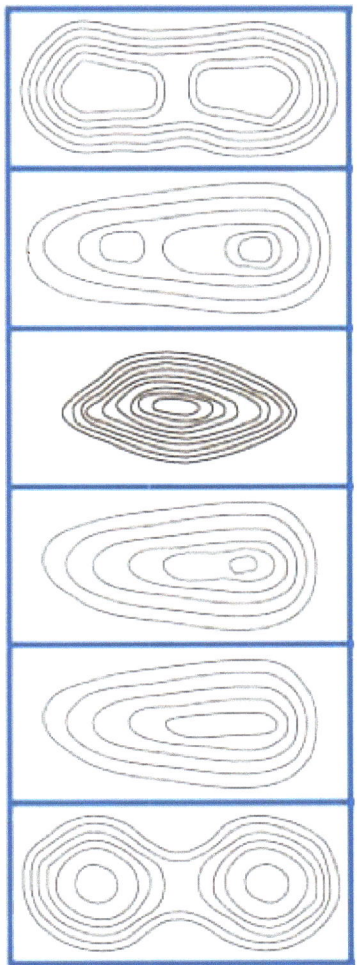

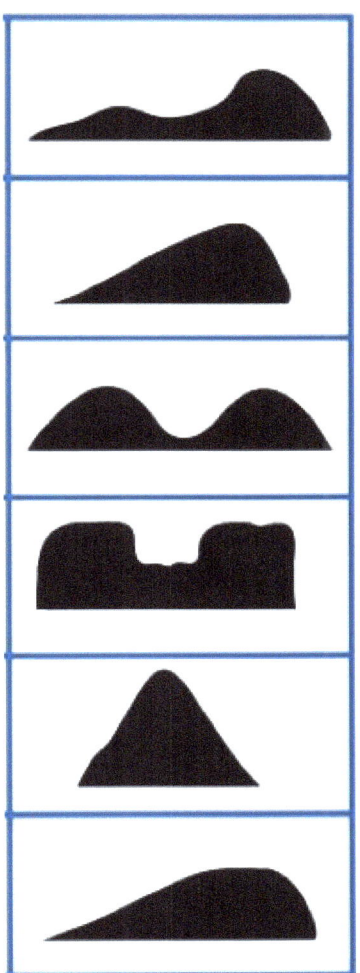

Draw a cross section (profile) of the hills on this map.

A vertical line has been drawn to help you identify where you should draw a high point. Draw in your own vertical lines to assist you in drawing the cross section.

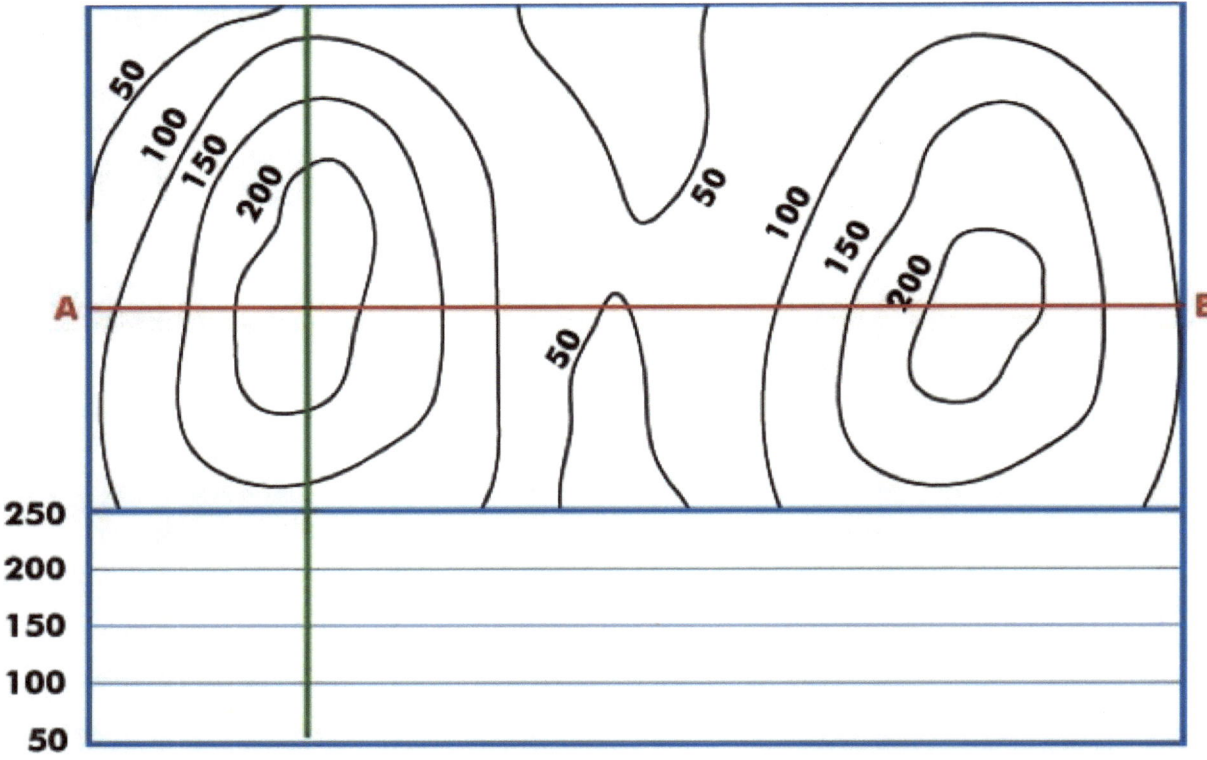

Contour Maps Quiz

1. A _____ map shows the surface features of an area using contour lines
    a. topographical
    b. political
    c. physical
    d. geographic

2. Contour lines never cross. True or false?

3. Contour lines that are close together indicate a steep slope. True or false?

4. The Great Plains have contour lines that are very close together. True or false?

5. Contour lines conner points of equal elevation. True or false?

Wind

Which runway and direction would your choose?

When planes take off, it is advantageous if they fly directly into a wind. If you know this fact, which runway would you choose, and in which direction would you take off?

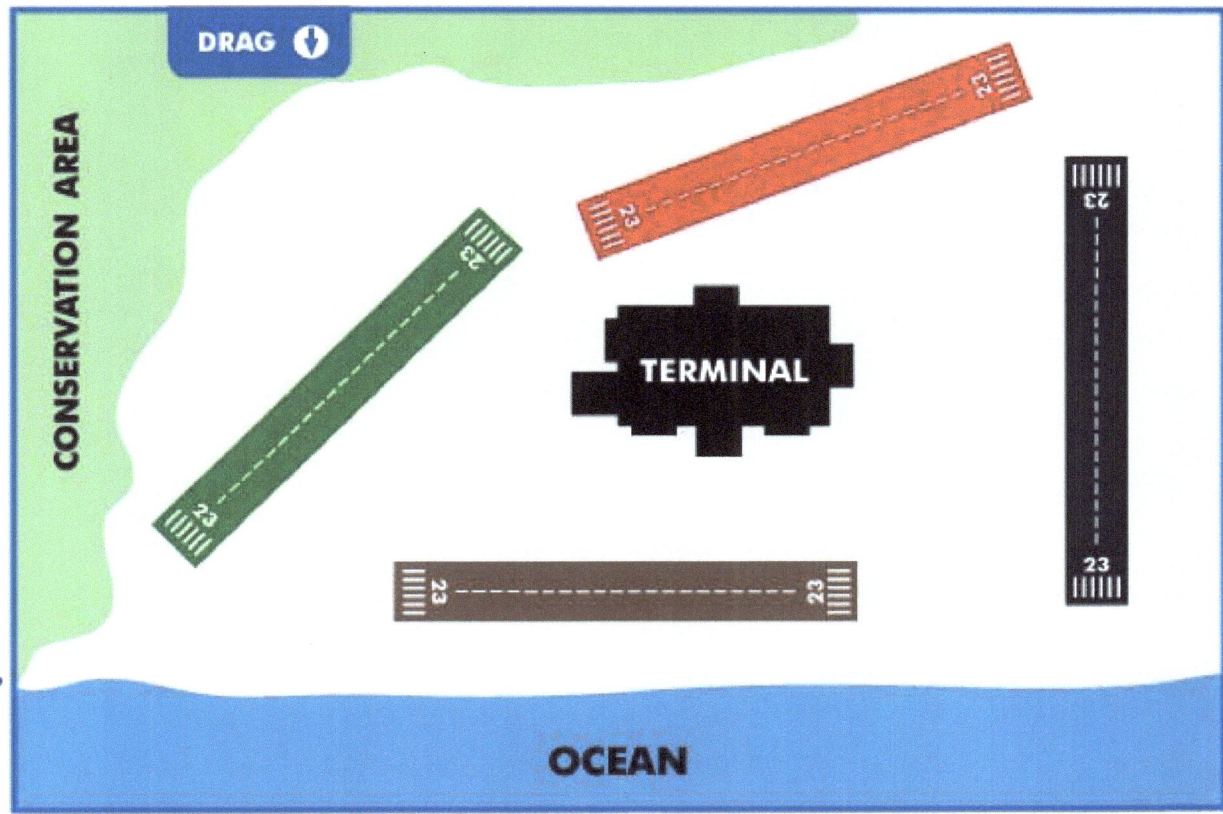

Draw the plane on the correct runway facing the correct direction.

During the daytime, the wind blows from the ocean to the land. In this lesson, we will be exploring how and why winds occur, and we'll find out why a daytime wind blows from the ocean to the land.

What is wind and what causes it?

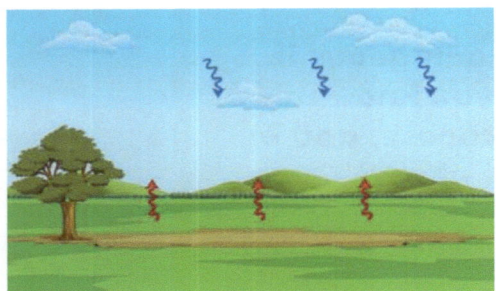

Winds are caused by the movement of air from areas of high pressure to areas of low pressure. But what are areas of high an flow pressure and why does wind move this way?

To answer both parts of this question let's start by recalling that Earth is shaped like a sphere and tilted on its imaginary axis. Because of this, certain parts of the Earth receive more direct sunlight than others. We can say that Earth is heated unequally.

The air above the earth is heated by heat that radiates or bounces off the surface. This air is also heated unequally.

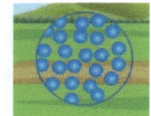 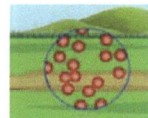

When radiated heat from the Earth's surface warms the air, the warm air expands and becomes less dense or lighter than cold air. That's why warm air rises. Cold air is more dense and heaver than warm air and so it sinks.

The terms high and low pressure refer to the density of the air in a given area. Cold air is more dense and so it exerts higher pressure. Warm air is less dense and so it exerts lower pressure. Air moves in a continuous cycle from areas of low pressure to areas of high pressure warm air rises and eventually cools and then sinks. Cool air eventually warms and rises. This movement of air from areas of high pressure to low pressure is what causes winds.

Other factors, that we will explore later, determine which way winds blow.

Sea and Land Breezes

Land absorbs and radiates heat more effectively than water.

During the day when the sun is out the land and the air above the land are warmer than the sea and the air above the sea.

Because of this temperature difference winds blow form the sea to the land. This is because the warm lower pressure air over land rises and then is replaced from the cooler air from the sea. We call this a sea breeze.

What do you think happens at night?

At night the opposite occurs. Winds blow from the land to the sea. This is because the water is warmer than the land at night. The warmer low pressure air above the sea at night rises and is replaced by the cooler air from land. This is called a land breeze.

Which runway and which direction?

Now that you understand winds a bit more and know that an airplane likes to take off into the wind, draw an airplane on the best runway for day and for night. Draw an arrow next to the runway to indicate wind direction.

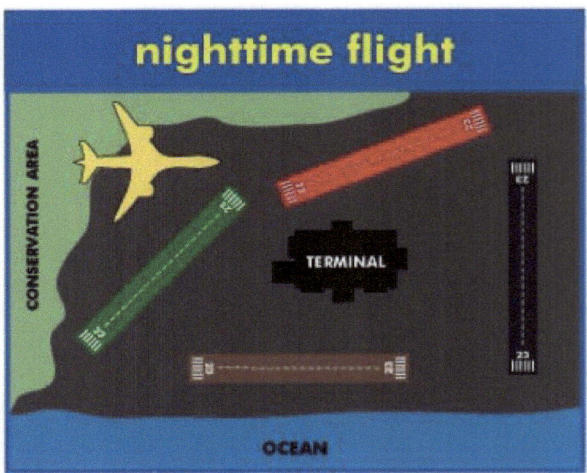

Local and Global Wind Patterns

Worldwide, we experience a number of different wind patterns. Wind patterns have both local and global causes. Local wind patterns like land and sea breezes occur in localized areas where land and sea temperatures are unequal.

Mountains and valleys also produce local winds due to unequal temperatures.

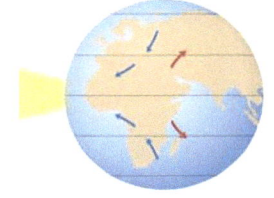

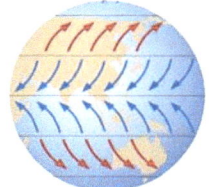

Global wind patterns also called wind belts occur as a result of low pressure at the equator. This is because the equator always gets more direct sunlight than the poles. As warm air rises from the equator toward the poles it then begins to cool.

However, because the Earth is rotating from west to east the winds curve as well. In the Northern Hemisphere they are deflected toward the right and in the Southern Hemisphere they are deflected toward the left.

Winds rising from the equator and toward the poles are known as the Prevailing Westerlies because the come from the west. They are the strongest winds and are responsible for many weather patterns and occur between 30° and 60° latitude in both hemispheres.

As warm air rises from the equator and toward the poles some of it cools and sinks back down toward the equator. This occurs at about 30° latitude in both hemispheres. These are known as the trade winds. Slow and steady winds that blow toward the equator from the NE and SE.

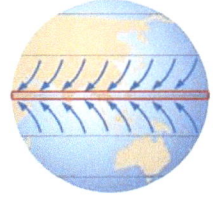

The Doldrums is the name given to a calm area just above and below the equator where the trade winds meet.

Meanwhile at the poles, cold air moves down toward the equator. These winds are known as the Polar Easterlies because they blow from east to west.

Label the Earth's major wind belts.

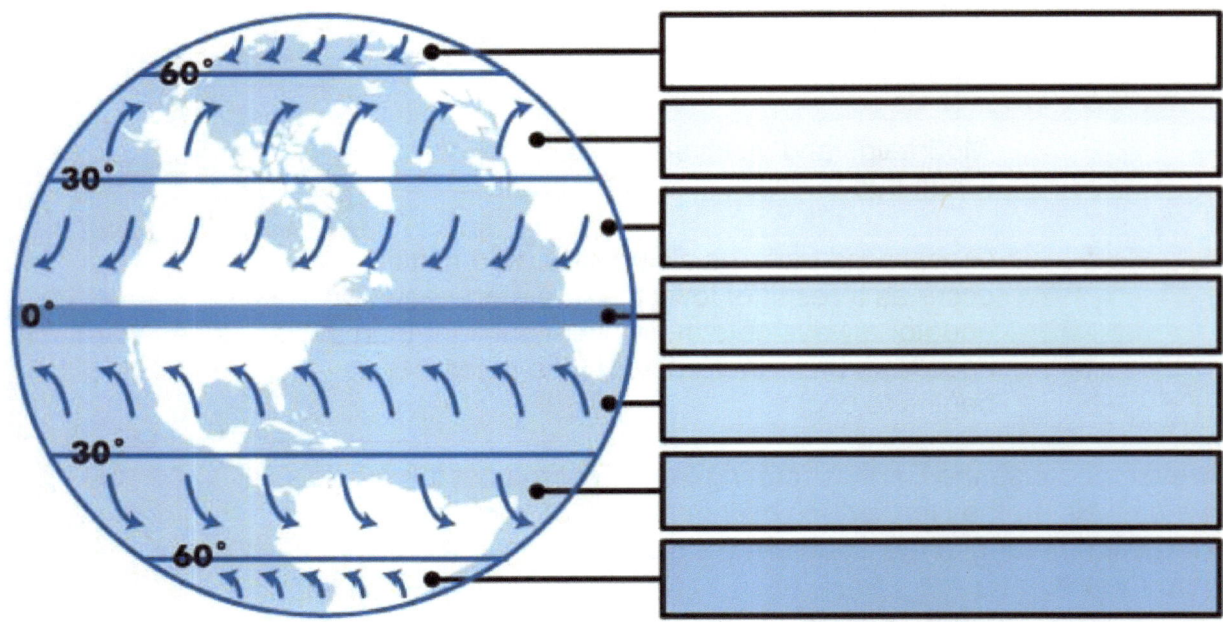

prevailing northwesterlies   **prevailing southwesterlies**

southeast tradewinds   **northeast tradewinds**   **doldrums**

prevailing southeasterlies   **prevailing northeasterlies**

Four tools that measure wind and atmospheric pressure.
Read the explanations of the different measuring device and then label the illustrations

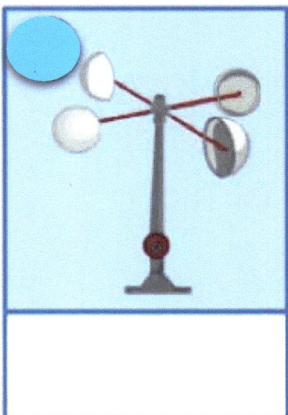 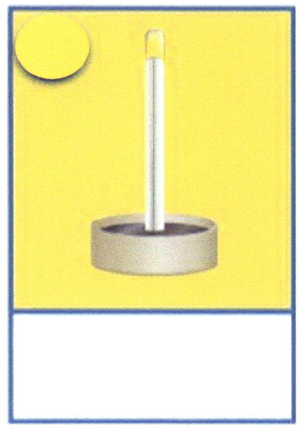  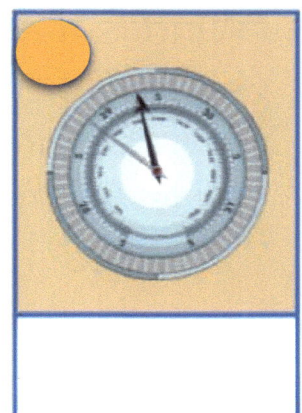

A weather vane is used to measure winds' direction.

It has a freely moving pointer seated above fixed markers; north, east, south, west.

Due to its design the arrow will point in the direction that the wind is coming.

For example if the wind is blowing north to south, the arrow will point toward the north.

An aneroid barometer is a watch like device that is used to measure atmospheric pressure.

An elastic disk surrounds a partial vacuum inside the barometer and is attached to a pointer.

When the disk expands or shrinks due to the change in atmospheric pressure, it moves the pointer.

An anemometer is a tool used to measure winds speed.

In the most basic design, four cups are attached to a fixed pole.

The wind speed is calculated by the number tomes the cups rotate around the pole one fixed amount of time.

A mercury barometer uses a glass tube filled with mercury to measure atmospheric pressure.

The glass tube stands upside down in a container called a reservoir also filled with mercury.

The mercury in the glass tube moves up and down as atmospheric pressure increases and decreases.

Name: _____

Wind Quiz

1. Wind is the movement of air _____.
   a. From an area of high pressure to one of low pressure.
   b. From an area of low pressure to one of high pressure.
   c. None of the above

2. The speed of wind depends on the pressure difference between warm and cool areas. True or false?

3. _____ is an instrument used to measure wind.
   a. Anemometer
   b. Hydrometer
   c. Chronometer

4. When air moves toward the equator, _____ bring rainstorms or cause cold conditions.
   a. westerlies
   b. easterlies
   c. trade winds

5. Where are the Doldrums located? _____

Newburyport, MA 01950

1-800-596-3175

OnBoard Academics employs teachers to make lessons for teachers! We create and publish a wide range of aligned lessons in math, science and ELA for use on most EdTech devices including whiteboard, tablets, computers and pdfs for printing.

All of our lessons are aligned to the common core, the Next Generation Science Standards and all state standards.

If you like our products please visit our website for information on individual lessons, teachers licenses, building licenses, district licenses and subscriptions.

Thank you for using OnBoard Academic products.